I0462350

DESIGN

SPECIE

Design Specification Concepts

FRANK APPIAH , research associate

KWAME NKRUMAH UNIVERSITY
OF SCIENCE AND TECHNOLOGY

KUMASI, ASHANTI, GHANA.

Author: Frank Appiah is known as an Ashanti-Ghanaian pioneer in computer engineering, software engineering and computing.

Frank Appiah hold studentships at the Bsc. (Bachelor in computer engineering, *Kwame Nkrumah University*) student Level in 2009, Msc.(software engineering, *King's College London*) student Level in 2009/10 and PGStd (post-graduate study, *King's/IEEE*) mission level in 2010. Student Appiah created and developed a software platform

called StreamEPS in the Java pro-
gramming language. Appiah is a
member of the National Academy
of Engineering in 2010/2011 and
holds the 'ING' title from Ghana In-
stitute of (Software) Engineering.

Preface

This book is a study on design specifi-cation and notation. This is the research outputs of the author's studies at Kwame Nkrumah University of Science and Technology, Faculty of Computer Engi-neering from 2009 to 2015. This book is a must read for all (post)graduate stu-dents, teachers, and more.

Table of Contents

1. *The Notion of Correctism in the Specification of Notation.*

2. *A Correct Specification Development in Engineering Design and Systems.*

Design Specification Concepts

Illustration Index

Illustration 1: Art Specification Diagram.........22
Illustration 2: Design Specification Diagram...45

The Notion of Correctism in the Specification of Notation.

(Version 1)

Design Specification Concepts

Abstract. A notation of art specification(spec) is provided in the teaching of correct-ism. The art specification diagram in this paper shows the properties of art specie of a real thing in a real world. The precise characteristics of production, thing, body and skill of art specie are enumerated.

Keywords. art spec; system science; specification; language; production; skill; body; thing;notation; art specie

1 INTRODUCTION

The notion of correct[1] is to make something (specie) right according to facts or rules. Otherwise , it is a notion of incorrect: That is to show that a specie is wrong. The notion of correct can make a specie work in the way that it should. The notion of correct-ism or correction-ism in the specification of notation is to inform a party that a specie of notation is said according to same facts or rules. The specification of notation becomes acceptable and corrective according to facts or rules. The notion of correct-ism takes on a way considering it's sociality and morality. The social and moral dimensions makes the specie of notation have a behavior. If be that a specie of notation is correct then it will be formal and correct. There can be some factual errors that needs correction in a specification[2] of notation. The corrective accords to facts or rules designs a behavior to solve a problem or improve a bad situation. The specification of notation is the specification of notations. The specification of notation is the cornerstone for the notion of correct-ism.

2 ART SPECIFICATION

The specification of notation corresponds to a real thing or object. The collaborate behavior created from the specification of notation is evident from the notation specie. The properties of an art specie of notations are:

- Production

- Thing

Design Specification Concepts

- Body

- Action/Skill.

It is proper in the production of an art specie to follow these precise characteristics;

(a) similar facts or rules to the real thing

(b) generalization of real object

(c) feature of real thing

(d) member of a thing

(e) group of real objects or happen objects.

Design Specification Concepts

The second property of an art specie of notation gives the characterization of :

1. Objective Specie

2. Possessive Specie

3. Activity or action of object specie

4. Event of Object-specie

5. Quality of Specie

6. Information of Species.

The art specie of notation demands that a specie has :

Design Specification Concepts

(a) Main (outer) part of thing

(b) Dead part of thing

(c) Drive part of thing

(d) Armless or legless part of thing

(e) Animate part of thing

(f) Appearance of thing

(g) Collection part of thing

(h) Main working of thing

(i) Grow part of thing

(j) Reproduce part of thing.

Finally, the *skill property* of art specification of notation will be characterized

by the skill setting of rules , skill of solving a feature problem to be a cornerstone of solution, skill of problem identification and skill of calculation. This develops a language for the characterization of the art specification of notation.

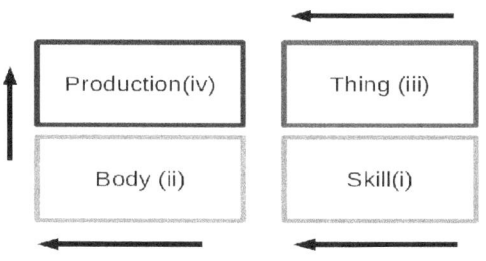

Illustration 1: Art Specification Diagram

Design Specification Concepts

The art of specification is a reverse process. A high production spec will lie on the skill horizon first, then on the body horizon secondly, thirdly will be the thing horizon. A vertical production spec depends on these three horizons. The assessment of a specification of notation requires a proper skill to develop character and behavior. What kind of skill is needed for an existence spec or non-existence spec? An assessment form demanding basic answers to a "Does a specification exists: Yes or No". The next question is "What kind of skill is needed for identifying problems before or after the spec?". An assessment form will ask if there is a sufficient (business) analyst to determine a solution to an identified problem. The form will also ask for solution accomplishment, solution appropriateness, solution limitation and solution usage control. The assessment of solution quality and solution rules or setting will be needed. A solution behavior is clearly described in spec notations. A solution behavior is described in spec notations but could use improvement. An assessment form will ask if there is a feature solution that clearly describes what a solution does and can a feature developer or skill solver implement the solution. An assessment form will determine the skill requirement for setting the rules of production, thing and body of rules that characterizes the spec notation.

The rules of production will allow the spec notation to grow and reproduce into a complex[3] set of rules. The rules of thing will help assess if the main objective specie, possessive specie, event of object specie, quality of specie etc are all met. The body of rules will help assess that the main (outer) part of thing, dead part of thing, drive part of thing, main working of thing etc are all met to the described spec notation. The body of rules will also take into consideration the body of standards in the spec notation of production service. An assessment form will determine the skill requirement for calculation that will provide a single, consistent way to access a responsible information to perform task. The performance of solution function will be determined by the skill of calculation and assessment of calculation. The security of solution function will be determined by the calculation assessments. An assessment form will ask of what critical value of each calculation assessment will or can affect the performance or security of a spec notation.

The semantic alignment characterizes the common understanding of calculation entities and solution task to provide the definitive source of information in each spec notation. The notion of correct is to make something (specie) right according to facts or rules of production, thing or body of rules.

3 CONCLUSION

The notion of making an art specie correct is introduced in section 1. According to facts or rules, one can make an art specie correct or incorrect. The proper working of a specie is the notion of correct specification. The specification of notation becomes acceptable and corrective according to facts or rules. In order to make a specie have a behavior, the social and moral dimensions need to be taken into consideration. The corrective accords to facts or rules designs a behavior to solve a problem or improve a bad situation. In improving a bad situation, the corrective accords to facts or rules is needed to solve design and behavior issues. The notation of specification corresponding to real things in real world is in section 2. The four properties of art specie is resulted from the art spec diagram namely production, thing, body and action or skill. The ten parts of thing in art specification is postulated. The development of the art specie language is the characterization of the art specification of notation. The art of specification is a reverse process.

REFERENCES

1. School Macmillan Dictionary, Correct language. Accessed on 2016.
2. Michael Rosen, Boris Lublinsky, Kevin T. Smith and Marc J. Balcer. Applied Service Oriented Architecture, Service-Oriented Architecture and Design Strategies, pg. 94. Wiley Publishing Inc. 2008.
3. M. Mitchell Waldrop. Complexity. Simon and Schuster Paperbacks. 1992.

A Correct Specification Development in Engineering Design and Systems.

(Version 2)

Design Specification Concepts

Abstract. A design notation of design specification (spec) is provided in the teaching of correct design specification. The design specification diagram in this paper shows the properties of design specie of a real thing in a real developing world. The precise characteristics of production, thing, body and skill of design specie are enumerated in this novelty paper.

> **Keywords**. design spec; system; spec-
> ification; language; production; skill;
> body; thing; design process; design
> specie; design theory; model

1 INTRODUCTION

The notion of correct[1] is to make design specie or theory right according to design facts or design rules in the design process. Otherwise , it is a notion of incorrect notation design: That is to show that a specie is wrong or bad. The notion of correct can make a specie work in the way that it should in the development process. The notion of correct design or correction-ism in the specification of design notation is to inform a design engineer or system developer or research engineer that a specie of design notation is said according to same design facts or design rules. The specification of design notation becomes acceptable and corrective according to design facts or design rules. The notion of correct design takes on a way considering it's sociality and morality. The social and moral dimensions makes the specie of design notation have a behavior. If be that a specie of design notation is correct then it will be formal and correct. There can be some design factual errors that needs correction in a specification[2] of design notation. The corrective accords to design facts or design rules designs a behavior to solve a problem or improve a bad situation. The specification of design is the specification of design notations. The specification of design notation is the cornerstone for the notion of correct design.

2 DESIGN SPECIFICATION

The specification of design notation corresponds to a real thing or object. The collaborate behavior created from the specification of design notation is evident from the design notation specie. The properties of a design theory of design procedure or step are:

- Production ,

- Thing ,

- Body and

- Action/Skill.

It is proper in the *production* of an design specie to follow these precise characteristics;

Design Specification Concepts

(a) similar design design facts or design rules to the real thing,

(b) generalization of real object ,

(c) feature of real thing,

(d)member of a thing and

(e) group of real objects or happen objects.

The second property of a process of design theory gives the characterization of :

1. Objective Specie,

2.   Possessive Specie,

3.   Activity or action of object specie,

4.   Event of object-specie,

5.   Quality of Specie and

6.   Information on Species.

The development of design specie or theory demands that a specie has :

(a) Main (outer) design of thing,

(b) Dead design of thing,

(c) Drive design of thing,

(d) Armless or legless design of thing,

(e) Animate design of thing,

(f) Appearance of thing ,

(g) Collect design of thing,

(h) Main working of thing,

(i) Grow design of thing and

(j) Reproduce design of thing.

Finally, the *skill property* of the process and development of design specification or design notation will be characterized by the skill setting by subject rules , skill of solving a feature problem to be a cornerstone of solution, skill of problem identification and skill of calculation or simulation. This develops a prototype language for the characterization of the design process of a design notation.

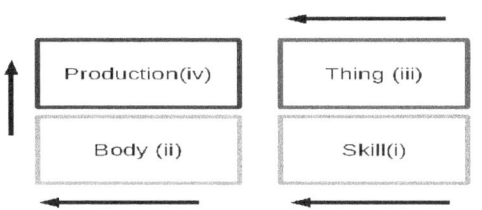

Illustration 2: Design Specification Diagram

Design Specification Concepts

The design of specification is a reverse process. A high production spec will lie on the skill horizon first, then on the body horizon secondly and thirdly will be the thing horizon. A vertical production spec depends on these three horizons. The development of prototype will be materialized at the skill and early production stage. The feature of new parts brings about the inventive or innovative design process. The assessment of a specification of design notation requires a proper skill in developing character and behavior of the engineer. What kind of skill is needed for an existence spec or non-existence spec? An assessment form demanding basic answers to a "Does a specification exists: Yes or No". The next question is "What kind of skill is needed for identifying problems before or after the spec?". An assessment form will ask if there is a sufficient (engineering) analyst or specification engineer to determine a solution to an identified problem. The form will also ask for solution accomplishment, solution appropriateness, solution limitation and solution usage control. The assessment of solution quality and solution rules or setting will be needed. A solution behavior is clearly described in spec design notations. A solution behavior is described in spec design notations but could use improvement. An assessment form will ask if there is a feature solution that clearly describes what a solution does and can a feature designer or developer or skilled engineer (solver) implement the solution. An assessment form will determine the skill requirement for setting the design rules of production, thing and body of rules that characterizes the spec design notation.

The design rules of production will allow the spec design notation to grow and reproduce into a complex[3] set of design rules. The design rules of thing will help assess if the main objective specie, possessive specie, event of object specie, quality of specie etc are all met. The body of design rules will help assess that the main (outer) part design of thing, dead part design of thing, drive part design of thing, main working development of thing etc are all met to the described spec design notation. The body of rules will also take into consideration the body of standards in the spec design notation of production service. An assessment form will determine the skill requirement for calculation that will provide a single, consistent way to access a responsible information to perform task. The performance of solution function will be determined by the skill of calculation and assessment of calculation. The security of solution function will be determined by the calculation assessments. An assessment form will ask of what critical value of each calculation assessment will or can affect the performance or security of a spec design notation.

The semantic alignment characterizes the common understanding of calculation or simulation[4] entities and solution task to provide the definitive source of information in each spec design notation. The notion of correct is to make something (specie) right according to design facts or design rules of production, thing or body of design rules.

3 RESULTS OF WORK

Design theory in engineering design procedure or process can take on the 4 principles of specification development

based on the *production* characterization, body design rules, *thing* processes
and *action* specification or *skill* (requirement) development. Design and development processes of engineering design
can now proceed with the production
characterization model, body-specie
characterization model, thing design
model and finally skill requirement
model.

4 CONCLUSION

The notion of making an design
specie correct is introduced in section 1.

Design Specification Concepts

According to design facts or rules, one can make an design specie correct or incorrect which will lead to right design theory or wrong design theory. The development of proper working of a specie is the notion of correct specification. The specification of design notation becomes acceptable and corrective according to design facts or design rules. In order to make a specie have a behavior, the social and moral dimensions need to be taken into consideration. The corrective accords to design facts or rules designs a behavior to solve a problem or improve a bad situation. In improving a bad situation, the corrective accords to design facts or design rules is needed to solve design and development behavior issues. The design notation of specification corresponding to real things in real world is in section 2. The four properties of design and development specie is resulted from the design spec diagram namely production, thing, body and action or skill. The ten part designs of thing in design specification is postulated. The development of the design specie language is the characterization of the design specification of design notation. The design of specification is a reverse process.

REFERENCES

4. School Macmillan Dictionary(2016), Correct language. Accessed on 2016.
5. Rosen M., Lublinsky B., Kevin Smith T. and Marc Balcer J.(2008). Applied Service Oriented Architecture, Service-Oriented Architecture and Design Strategies, pg. 94. Wiley Publishing Inc.
6. Mitchell Waldrop M.(1992). Complexity. Simon and Schuster Paperbacks.
7. Gould H. / Tobochnik J. (1938), An Introduction to Computer Simulation Methods Part 2, Addison Wesley Publishing Company.
8. School Macmillan Dictionary(2018), Design Language. Accessed on 2018.

Design Specification Concepts

Alphabetical Index

A
Action..33, 42, 44
Activity...33
Animate...34
Appearance..34
Armless...34
Assessment..40
B
Behavior..43
Body..42, 44
C
Calculation..40
Characteristics...10
Characterization....................................42, 44
Characterization model...............................42
Collect..34
Complex..39
Control..37
Corrective..44
D
Dead...39
Dead design..33

Design Specification Concepts

Design..42, 44, 45
Design procedure..41
Design rules..41
Design theory...41
Designs..43
Development..44
Diagram...44
Dimensions..43
Drive...34
E
Engineer...38
Event..33, 39
F
Facts...43
Feature..32
Function...40
G
Generalization..32
Group..32
H
High production..36
I
Information...33, 41
L
Language..44
M

Design Specification Concepts

Main...33
Main working..34
Member..32
Model...42
Moral..43
Moral dimensions..29
N
Notation..44
Notion..42
O
Objective..39
Objective Specie...32
P
Part..44
Possessive..39
Possessive Specie...33
Process...36, 41, 45
Production..36, 42, 44
Properties...44
Q
Quality..33, 39
R
Real thing...32
Real world..10
Requirement...42
Reverse...36, 45

Design Specification Concepts

Rules...43
S
Second property...32
Security..40
Semantic..40
Similar design..32
Simulation..41
Single...40
Skill..44
Social...43
Solution..37, 40
Solver...38
Specification...10, 45
T
Thing...42, 44
Thing design..42
V
Vertical...36

www.ingramcontent.com/pod-product-compliance
Lightning Source LLC
Chambersburg PA
CBHW061225180526
45170CB00003B/1162